Azubuike Chidowe Odunze

Conservation Agriculture: A Nigerian Synthesis and Mitigation Measures

Azubuike Chidowe Odunze

Conservation Agriculture: A Nigerian Synthesis and Mitigation Measures

Conservation Agriculture: A Nigerian Synthesis and Mitigation Measures

LAP LAMBERT Academic Publishing

Impressum / Imprint
Bibliografische Information der Deutschen Nationalbibliothek: Die Deutsche Nationalbibliothek verzeichnet diese Publikation in der Deutschen Nationalbibliografie; detaillierte bibliografische Daten sind im Internet über http://dnb.d-nb.de abrufbar.
Alle in diesem Buch genannten Marken und Produktnamen unterliegen warenzeichen-, marken- oder patentrechtlichem Schutz bzw. sind Warenzeichen oder eingetragene Warenzeichen der jeweiligen Inhaber. Die Wiedergabe von Marken, Produktnamen, Gebrauchsnamen, Handelsnamen, Warenbezeichnungen u.s.w. in diesem Werk berechtigt auch ohne besondere Kennzeichnung nicht zu der Annahme, dass solche Namen im Sinne der Warenzeichen- und Markenschutzgesetzgebung als frei zu betrachten wären und daher von jedermann benutzt werden dürften.

Bibliographic information published by the Deutsche Nationalbibliothek: The Deutsche Nationalbibliothek lists this publication in the Deutsche Nationalbibliografie; detailed bibliographic data are available in the Internet at http://dnb.d-nb.de.
Any brand names and product names mentioned in this book are subject to trademark, brand or patent protection and are trademarks or registered trademarks of their respective holders. The use of brand names, product names, common names, trade names, product descriptions etc. even without a particular marking in this work is in no way to be construed to mean that such names may be regarded as unrestricted in respect of trademark and brand protection legislation and could thus be used by anyone.

Coverbild / Cover image: www.ingimage.com

Verlag / Publisher:
LAP LAMBERT Academic Publishing
ist ein Imprint der / is a trademark of
OmniScriptum GmbH & Co. KG
Heinrich-Böcking-Str. 6-8, 66121 Saarbrücken, Deutschland / Germany
Email: info@lap-publishing.com

Herstellung: siehe letzte Seite /
Printed at: see last page
ISBN: 978-3-659-79206-9

Copyright © 2015 OmniScriptum GmbH & Co. KG
Alle Rechte vorbehalten. / All rights reserved. Saarbrücken 2015

CONSERVATION AGRICULTURE: THE NIGERIAN SYNTHESIS AND MITIGATION MEASURES

By

Odunze A. C. (PhD)
Specialist Soil & Water Conservation/Land use Management

Department of Soil Science/Institute for Agricultural Research, Faculty of Agriculture, Ahmadu Bello University, Zaria Nigeria.
E-mail: odunzeac@yahoo.com odunzeac@gmail.com
Phone: +2348035722052

Up-down slope ridging impoverishes soil   Land preparation removing crop residues

EXECUTIVE SUMMARY:

Intensive agriculture, poor soil management practices by farmers and other soil users, and overgrazing aggravates soil degradation; such that the commonly resulting visible indices of degradation is soil erosion (by wind and water) and reduced crop yields. The negative impact of soil degradation due to inappropriate land management have become very apparent in Africa and Nigeria in particular; having severe economic, social and environmental effect on the populace. In order to mitigate effects of soil degradation by restoring the soils' quality and fertility to enhance farmer livelihoods and food security, Conservation Agriculture strategies are advocated for Nigerian agriculture. Conservation Agriculture is a concept for natural resource-saving, strives to achieve acceptable profits with high and sustained production levels while concurrently conserving the environment. In the humid zones of Nigeria, intensive/continuous cultivation of available land using modern farming equipment has increased soil disturbance and removal of soil cover to accelerate soil erosion in these humid zones. In particular, slopping lands that would not normally be put into regular cultivation are being intensively cultivated for crops like maize, yam and cassava without using appropriate conservation measure and gullying has occurred in several locations to degrade land in the zone. In the sub-humid zones, land preparation generally begins from the dry months of April while rainfall often establishes in the month of June with sporadic rains received in March, April and May. Wind pressure in these months cause huge soil materials to be lost to wind erosion, giving rise to 'dune sand' phenomenon commonly reported in the zone. The implication of this is that land preparation in the zone is done when the soil is not sufficiently moist or lacks sufficient water to resist erosion by wind and water. Conservation agriculture which aims at enabling farming communities develop agricultural practices that maximizes economic returns from the land while ensuring that the soil is not degraded in particular by erosion or agro-chemicals use in the process of farming is advocated. Conservation agriculture practice for each designated community should be well researched and packaged for farmer adoption under a researcher-farmer participatory arrangement. Government intervention with community level planned whole farm programme to enable farming communities develop conservation agriculture practices that maximizes economic returns from the land while ensuring that the soil is not degraded in particular by erosion or agro-chemicals use in the process of farming and provision of grants, insurance covers and aids for conservation agriculture farmers is strongly advocated.

CONTENTS	Pages
Title page	1
EXECUTIVE SUMMARY	2
1.0 INTRODUCTION	5
2.0 STATUS OF CONSERVATION AGRICULTURE IN NIGERIA	6
2.1 Country Background	6
2.2 Population	6
2.3 Weather	6
3.0 HUMID ZONE ON-FARM CONSERVATION AGRICULTURE ACTIVITIES	7
3.1 Activities	7
3.1.1 Land preparation	7
3.1.2 Fertilizer application	8
3.1.3 Humid zone conservation Agriculture land use Management	9
3.1.4 Stakeholders and support	9
3.1.5 Lessons learnt	10
3.2 SUB-HUMID ZONE ON-FARM CONSERVATION AGRICULTURE PRACTICES	10
3.2.1 Land preparation	10
3.2.2 Fertilizer application	12
4.0 CONSERVATION AGRICULTURE LAND USE MANAGEMENT	13
4.1 Strip cropping	14
4.2 Short fallows	16
4.3 Mono-cropping	17
4.4 Tied ridging	18
4.5 Planting pits	20
4.6 Stone-bench or mound terracing	20
4.7 Cover cropping/Mulch practices	22
4.8 Residual moisture harvesting	24
4.9 Irrigation/supplemental irrigation	24
5.0 LIVESTOCK FARMING	28
5.1 Restricted grazing and paddock ownership	29
5.2 Crop-livestock integration	30
5.3 Organized grazing land	31
5.4 Lessons learnt	31
5.5 Stakeholders and support	32
6.0 MAIN CONSTRAINTS TO DEVELOPMENT OF CONSERVATION AGRICULTURE IN NIGERIA	32
6.1 Government policy	33

6.2 Empowered Research and Extension outfit	33
6.3 Farmer Empowerment	34
6.4 Suggested Areas Of Research Focus	34
6.5 Conclusion	35
6.6 References Consulted	35
7.0 SUGGESTED CONSERVATION AGRICULTURE PRACTICES FOR RICE AND CASSAVA PRODUCTION IN NIGERIA	38
7.1 Soil fertility and quality management for rice (*Oryza sativa* L) production	38
7.1.1 Site selection and land preparation	38
7.1.2 Varieties	39
7.1.3 Weed Control	40
7 1.4 Cultural practices	41
7.1.5 Fertilizer use, type and rate	41
7.1.6 Nutrient deficiency/toxicity symptoms	43
7.1.7 Pest and disease incidence	45
7.1.8 Yield Expectancy	45
7.1.9 Agricultural Extension Agents (EAs)	45
7.1.10 References consulted	45
7.2 SOIL QUALITY AND FERTILITY MANAGEMENT FOR CASSAVA PRODUCTION UNDER CONSERVATION AGRICULTURE	46
7 2.1 Introduction	46
7.2.2 Site selection	47
7.2.3 Varieties	47
7.2.4 Weed control and land preparation	48
7.2.5 Planting and planting material	48
7.2.6 Fertilizer use, type and rate	49
7.2.7 Yield Expectancy	49
7.2.8 Agricultural Extension Agents (EAs)	50
7.2.9 References consulted	51

1.0 INTRODUCTION

Everywhere in the world where people change a natural ecosystem into agriculture, the land degrades. As the land become increasingly degraded and thus less productive, farmers are forced to further overuse the land. The intensive agriculture and overgrazing that follows further aggravates soil degradation. The visible part is erosion; when soil particles leave the land, mobilized and deposited by gravity, water or wind. Soil erosion can therefore be seen as both a symptom of underdevelopment (i.e. poverty, inequality and exploitation), and as a cause of underdevelopment. Some erosion is however natural but present rates are worrying as it is significantly accelerated mostly by inappropriate land use activities associated with agriculture. Worldwide, soil erosion occurs and rates as high as 390 t/ha/year has been recorded in the hilly area of Loess Plateau of China (Inman, 2006). In Nigeria, Odunze (2003) stated that potential soil loss at 4 % slope in the Guinea Savanna of Nigeria was between 11.4 and 13.34 t/ha/year and averaged about 12.37 t/ha/year under erosion caused by water in Zaria area. Soil loss under wind erosion is a common occurrence in the arid and semi-arid Africa, suggesting that soil loss values in sub humid Africa and Nigeria in particular would be greater than 12 t/ha/year and soil degradation by this erosion phenomenon would have huge economic and social implications on the Nigeria populace. The negative impact of soil erosion due to inappropriate land management have become increasingly apparent in Africa and Nigeria in particular; having severe economic, social and environmental effect on the populace. A number of factors are responsible for this increase and include crop and livestock production on inappropriate land, overstocking and overgrazing of land, wrong timing of agricultural practices, uncontrolled tree felling and lack of timely good ground cover (Inman, 2006). For example, it is a common sight for crops like maize to be grown on slopping land with ridges made on up-down slope direction. In this instance, furrows would conduct runoff from the fields at erosive rates to further degrade the land with surface wash. Crop harvest is done between the months of November and January in the sub humid zones of Nigeria and livestock is left to graze fields without control. The consequence of this is that animal hoofs disturb the land to destroy soil structure, all available grazing material is removed from soil surface and the soil is left bare of any vegetative cover from December to May. During this period wind speed is high and dry, and soil materials are lost to wind erosion, leaving the soil further degraded before the subsequent cropping season. Fortunately there are many ways to reduce erosion (Anthoni, 2000). However, the above discussion presents a cyclic soil degradation picture for the Nigerian agricultural soils and need to be reversed by use of appropriate conservation agriculture practices.

2.0 STATUS OF CONSERVATION AGRICULTURE IN NIGERIA:
2.1 Country Background
Location

Nigeria is located in West Africa on the coast of Atlantic Ocean. The Country is bordered by Cameroon Republic in the East; Niger and Chad Republics in the North and North east, respectively; Benin Republic in the West; and Atlantic Ocean in the South (Fig.1). Nigeria has a total land area of 923,773 square kilometers and is richly endowed with abundant and diverse natural resources, both renewable and non-renewable (Shonekan, 1997).

Fig.1 Map of Nigeria showing States and the Federal Capital Territory

2.2 Population
Nigeria is the most populous nation in Africa and the 11th in the World and has a population of 150 million based on 2006 National Population and Housing Census. About 45% of the population is below the age of 15 years, while only 3 per cent is above 64 years.

2.3 Weather

In almost all geographical zones of Nigeria, the wet and dry seasons are fairly well marked. The mean air temperatures are relatively constant throughout the year, a favourable situation for growing almost any crop in the country. Variation in air temperature follows altitude and proximity to large water bodies such as Atlantic Ocean.

Conservation agriculture in this discussion would therefore consider On-farm conservation activities in Nigeria. For the purpose of this presentation also, conservation agriculture in Nigeria would be sectioned into Humid and Sub humid zone conservation agriculture practices. Conservation agriculture in this presentation would thus be defined as those developed and adopted On-farm activities that aim at enabling farming communities maximize economic returns from the land while ensuring that the soil is not degraded; in particular, by erosion or agro-chemicals use in the process of farming. Conservation Agriculture however implies agricultural practices that employ minimum soil disturbance; such as zero tillage, reduced tillage practices, crop rotation practices, permanent soil cover in such a way that optimal agricultural production is attained while the soil conditions are improved and conserved. Therefore, if conservation agriculture is implemented well in Nigerian agricultural systems, it would improve efficiency of inputs, increase farm income, improve and/or sustain crop yields, protect and revitalize soil, biodiversity and the natural resource base. Conservation Agriculture is a concept for natural resource-saving, strives to achieve acceptable profits with high and sustained production levels while concurrently conserving the environment (FAO, 2009).

3.0 HUMID ZONE ON-FARM CONSERVATION AGRICULTURE (CA) ACTIVITIES
3.1 Activities:
3.1.1 Land Preparation
Crop production activities in the humid zone of Nigeria begin with land preparation. The zone experiences bimodal rainfall with annual rainfall amount often exceeding 2000 mm, thus supporting two cropping seasons in any one year. The first cropping season begins from about late January with bush clearing and clustered-trash burning. The second cropping season begins from about September and does not necessarily involve bush clearing as planting is commonly done in farms where crops like maize has been harvested. At land preparation, bush is not completely cleared of trees and shrubs. Tree branches are pruned and young shrubs and grasses

are cut, left to dry, gathered in several heaps and subsequently burnt. This practice could be improved upon to aid check soil erosion and enrich soil carbon by sorting to remove cut woody materials that could obstruct cultivation activities and plough into the soil all decomposable plant and animal materials.

Commonly, mounds are made such that water is accommodated in small pockets left around the mounds while crops are planted in the mounds. On slopping lands, this practice has proved disappointing as the pockets soon exceed their capacity to store water and runoff ensues. Consequently rilling, channeling and gullying commence as soil carrying capacity of the runoff increase with increasing volume of runoff water and length of slope of the farm area. This has been blamed to have caused most erosion problems in the densely forested South-Eastern States of Nigeria (Dale, 1994). Ridging has also been adopted by a few farmers in the humid zones of Nigeria, but they are yet to adopt terracing, contour bund and contour ridging practices that would make ridge farming anti-erosive especially in slopping lands. The general practice of burning plant materials at land preparation and not incorporating organic matter renders the soils very susceptible to erosion.

3.1.2 Fertilizer application:

Soil erosion occurs when the soil is largely bare of vegetative cover (Odunze, 2003, Lal, 1975). Therefore between seeding and crop establishment maximum soil loss is witnessed under rain-fed crop production. Also fertilizer is applied to the soils within this period and is lost to erosion, implying wasted effort and finance, with attendant soil degradation. Worse still, farmers in the humid zone of Nigeria scarcely use inorganic fertilizer materials for crop production. When they do, grossly insufficient quantity is applied to the soil. Therefore the soil is not optimally nourished and good ground cover is not often rapidly generated to shield the soil surface from effect of splash erosion that detaches, dislodges and transports soil materials away from its original location. Contour farming practices that would ensure that soil and water are in place to retain applied fertilizer materials for plant roots absorption and excess field water is drained without causing erosion is advocated for this zone. Also small ruminants herding and poultry farming within homesteads are common livestock farming practices in the zone but no organized plan is noticed to return farmyard manure, feed remains and urine to the soil, nor is compost making a popular practice in the South-Eastern States of Nigeria. The implication is that soils in cultivable areas have low organic matter contents and a dominance of low activity clays like kaolinite, to be very susceptible to erosion by water.

3.1.3 Humid zone Conservation Agricultural Land use Management:

Shifting cultivation used to be capable of restoring fertility of agricultural lands in the humid zones of Nigeria. Then, cultivation was limited to the use of hoes and matchets with minimal soil disturbance, and farming could continue on the land until yield return from the soil is noticed to be discouraging. Thereafter, the farmer moves to another piece of land and allows the previous land to regain fertility over a period of about five years or more. Common crops grown in this zone include yam, cassava, maize, cocoyam and vegetables. Presently, human population increases has put much pressure on land for such other uses as residential, industrial, transport and other commercial uses to the disadvantage of sufficient land for crop production under shifting cultivation practice. Therefore, intensive/continuous cultivation of available land using modern farming equipment has increased soil disturbance and removal of soil cover to accelerate soil erosion in these humid zones. In particular, slopping lands that would not normally be put into regular cultivation without proper conservation measures are being intensively cultivated for crops like maize, yam and cassava without using appropriate conservation measure and gullying has occurred in several locations to degrade the land in the zone.

3.1.4 Stakeholders and Support:

Some Conservation Agriculture Stakeholders in the humid zone of Nigeria include agricultural research institutes like the Federal Universities in the zone; with a few of these: Federal Universities of Agriculture, e.g., Michael Okpara University of Agriculture, Umudike, Umuahia; University of Nigeria Nsukka, Nigeria; Unversity of Calabar, Calabar; Abia State University, Obafemi Awolowo University Ile-Ife, University of Ibadan, University of Agriculture Abeokuta, Institutes like International Institute of Tropical Agriculture, International livestock Research Institute, National Roots Crop Research Institute Umudike, Umuahia, Institute of Agricultural Research and Training, Ibadan, Colleges of Agriculture in Nigeria, Federal and State Ministries of Agriculture and Agricultural Development Projects in all Southern (Fig.1) parts of Nigeria, farmers, Agricultural support banks and the private sector

investing on agricultural products (for exports, industrial, and merchandise). Supports from these organizations include:
- Technology acquisition and transfer
- Extension message delivery and farmer education
- Micro finance supports
- Market Information provision
- Agricultural Education

3.1.5 Lessons Learnt:
Among lessons learnt in agriculture as practiced in the humid zone of Nigeria include the following:
- At land preparation, cut plant materials are burnt in clusters
- Slopping lands that would ordinarily not be cultivated regularly without adequate conservation measures are intensively cultivated with improper conservation measures
- Organic and inorganic fertilizer materials are not adequately used for crop production
- Soil conservation support practices like terracing and contour ridging are not commonly practiced
- Human population increases in the zone has necessitated intensive cultivation of arable lands as available land is also put into residential, industrial, commercial and transport uses
- Relevant Stakeholder institutions to support conservation agriculture in the zone exist

3.2 Sub-Humid Zone On-Farm Conservation Agriculture Practices:
3.2.1 Land Preparation
In the sub-humid zones of Nigeria, land preparation generally begins from the dry months of April. Rainfall often establish in the month of June while sporadic rains may be received in the months of March, April and May. Wind pressure in these months cause huge soil materials to be lost to wind erosion, giving rise to 'dune sand' phenomenon commonly reported in the zone. Rainfall amounts in this zone often does not exceed 1200 mm/annum but could be less than 600 mm/an in some ecologies and years. The implication of this is that land preparation in the zone is done when the soil is not sufficiently moist or lacks sufficient water to resist erosion by wind.

Also, organic matter from homesteads (Plate. 1) is randomly distributed in farms (Plate. 2) from about March to await land preparation when they would be incorporated into the soil. Organic matter so distributed would largely be lost to wind erosion; thus polluting the atmosphere, before being incorporated. Composting of organic matter and conveyance to farms at land preparation is advocated.

Plate 1: Organic matter gathering from homesteads

Plate 2: Organic Matter distribution on Farms

Land preparations in most places require ridging though planting on the flat is also practiced. However, farmers have not adopted contour farming practices; perhaps because they are not aware of the gains of contour bunding and ridging of their farms, to ensure sustainable conservation management of soils and enhanced economic yield returns from their farm activities. In consequence, ridging along slope direction is common (Plate 3) and soil erosion on farm land occurs and most farmlands have been degraded or out rightly destroyed by gully erosion.

Plate 3: Showing ridging along slope direction

Crops grown in this zone include maize, sorghum, millet, cowpea, soybean, groundnut, cotton and vegetables. Planting begins from about April (dry planting) and could continue to August (depending on the crop and purpose). Early (April) planting is done as security against short rain fall duration, low rainfall amount and dry spells, but often dune sand materials bury the seeds or seedlings so deep that germination is very low and farmers are compelled to replant. Also, early planted crops often fail due to the occurrence of dry spells that could last weeks and into critical phonological stages of crop growth. Therefore, recommended optimal land preparation time in the zone is the month of June when the soil is moist (Odunze *et al.*, 1997; Odunze and Kure, 2007; Odunze, 2011; Odunze, 2015), and planting could follow after this. This practice ensures that soil is not eroded at land preparation and optimal land management is put in place before planting, fertilizer applied is in the soil for crop roots uptake and erosion does not occur to degrade the soil before sufficient canopy/ground cover is established to shield off splash erosion or surface wash on the fields. In slopping lands and Plateaus (Jos and Mambila areas) in Nigeria, hill slope farming is common and terracing and ridging against slope direction is practiced in such areas (Longtau *et al.*, 2002). However, in such areas, expert assistance would be necessary to aid optimal natural resource conservation and management, for optimal economic yield returns from farm business/ investments.

3.2.2 Fertilizer application:

Soils in the sub humid zones of Nigeria have low organic carbon, low phosphorus, nitrogen and cation exchange capacity and are said to be inherently low in fertility status (Jones and Wild, 1975; Lombin, 1987; Odunze *et al* 1987; Odunze *et al.*, 2013). Crop and fodder production in the area is done with fertilizer application (organic and /or inorganic) to improve soil quality, fertility and boost soil nutrient release potential for roots uptake and enhanced crop yield. However, wrong methods, rates and timing of fertilizer application has led to soil degradation and yield impoverishment in some places. For example, farmers in the arid and semi arid regions of Nigeria apply fertilizer materials on the soil surface, hoping that it will be absorbed into the soil. This practice though easy and saves cost, is not efficient as the fertilizer materials get lost to soil erosion and volatilization. Sediments so carried would be deposited in natural drains (Water bodies) to pollute surface and underground water, endanger aquatic life and contribute to global warming by the greenhouse gas emissions consequent on this practice. Conservation agriculture practice for optimal fertilizer application method is to bury applied fertilizer material 10 cm deep and 10 cm away from crop roots (Enwezor *et al.*, 1989; Odunze and Ogunwole, 2002).

Also, fertilizer application rates for crops were developed by scientist (Enwezor *et al.*, 1989) but have not been well adopted by farmers. In consequence, some farmers under apply while others over apply fertilizer materials to the soil and the result has been that of adverse soil mining by crops (in the case of under application) and soil acidification, pollution of water bodies and salinization (in the case of over dosing of fertilizer) of the soils with attendant yield reduction, farmer impoverishment and environmental degradation. Timing of fertilizer application is also crucial to conservation agriculture; especially realizing the pattern of rainfall and seedling growth of crops. While some farmers apply fertilizer to crops up to four times in some instance, fertilizer application for sustainable conservation agriculture in the arid and semi arid zones of Nigeria is twice in the season (either at land preparation or at first weeding and just before crop flowering stage). This practice ensures that enough moisture is available in the soil to dissolve applied fertilizer materials and make them available to crop root and not to cause salinity or degrade the soil, water bodies or aquatic life.

4.0 CONSERVATION AGRICULTURE LAND USE MANAGEMENT:
Conservation agricultural land use practices in the sub humid zones of Nigeria include

i). Strip cropping
ii). Short fallows
iii) Cover cropping/mulch cropping
iv) Mono Cropping of cereals
v). Tied ridging
v). Planting pits
vi). Stone or Mound Terracing
vii). Residual Moisture Harvesting
ix). Irrigated/Supplemental irrigation
x). Crop/Livestock farming

4.1 Strip cropping:

Strip cropping practice is not limited to sloping lands where its importance lies largely in reducing slope length and soil carrying capacity of runoff, increasing infiltration rates of water and controlling erosion. In the sub humid zone of Nigeria farmers plant crops in strips to guard against crop failure, given the uncertain rainfall distribution pattern in the area. However, the practice is also of importance to conservation agriculture because in the practice two rows of cereal crops could be followed by four rows of legumes (Plates 4a, b & c) for example. The rows of legume crops improve the soils' quality and fertility even in the current year. In subsequent year, the two rows of previous year cereals and the two rows planted to legumes in previous year would be planted with new four rows for legumes. Such a sequence ensures that agricultural soils are improved annually and presents a crop rotation picture though within a reduced space in the farm. It allows a farmer to obtain a minimum of two crops in an organized manner in a farm while concurrently conserving the soil for sustainable production. The practice has found conservation agriculture use also for control of obnoxious weeds such as *Striga hermonthica* and soil improvement when legumes having striga control mechanisms are used. Striga infestation has caused farmers to abandon farm lands in the sub humid savanna of Nigeria (Emechebe *et al.*, 2003) and strip cropping with appropriate legumes have restored the soils for sustainable productivity. Such legumes that have proved effective for soil quality and fertility improvement and Striga control in the Savanna zones of Nigeria include *Desmodium uncinatum, Desmodium intortum, Macrotyloma uniflorum*, Soybeans *(TGX1448-2E and TGX 1951), some Cowpea varieties* and *Centrosema pascuorum*.

Plate 4a: Showing strip cropping using soybean (TGX 1448-2E)

Plate 4b: Showing strip cropping practice using *Macotyloma uniflorum*

Plate 4c: Showing strip cropping using Cowpea

4.1 Short fallows:

Short fallows involve planting of sole herbaceous legumes for the purpose of improving the soil for not more than two years. Thereafter, crop residues from the field would either be incorporated into the soils to grow cereal crops or cereal crops would be planted in the field under a zero till practice. Fodder legumes like *Centosema pascuorum*, *Lablab purpureus*, *Desmodium uncinatum* and *Macrotyloma uniflorum* that will provide quick ground cover and food legumes like cowpea, groundnut, and soy beans are commonly used. Under this practice Tarawali (1991) and Odunze *et al.* (2003) have shown that 60 % nitrogen needed to grow maize crop was obtained, and soil organic matter content contribution of the practice doubled, while soil erosion was controlled in fields put under short fallow practice. Farmers in the Northern Guinea Savanna of Nigeria have used this practice to reduce inorganic fertilizer application for sorghum and obtained good crop yield (Tarawali *et al.*, 2003). The practice however improves structure, quality, organic matter, fertility and nutrient mineral contents in the soil. Also the practice has been used in Nigeria to control obnoxious weeds such as *Imperata cylinderica* and *Striga species* (Plates 5a & 5b). In this context, *Mucuna pruriens* was planted at 30 cm interplant distance after land preparation for two years during which period Mucuna sufficiently smothered *Imperata cyliderica* to reduce weeding pressure for subsequent crop production. In the case of cowpea and soybean, varieties such as IT93k-452-1 and TGX-1448-2E (respectively) having striga suppression capacity has been used in Nigeria to reduce striga seed bank, cause suicidal striga seedling germination and improve quality fertility of soils for sustainable crop and livestock feed production.

Plate 5a: Showing short fallow with cowpea crop

Plate 5b: Showing Mucuna short fallow suppressing *Impereata cylinderica* Weed

4.3 Mono Cropping:

Mono cropping of cereals or legumes is also practiced in the sub humid savannas of Nigeria. Though the practice could give initial good yield, its drawback is largely that of soil erosion acceleration and soil quality degradation, in particular for cereal crops. Cereal crops like maize, sorghum and millet are planted sole in fields either on ridges or flats. Either way, root arrangement of these crops allow soil covering or between roots to be easily eroded to expose crop roots and sub-surface soil materials. Ridging in the area does not necessarily follow contour lines; hence the ridges serve to conduct water at erosive rates in the fields and rapidly degrade agricultural lands (Plates 6a and 6b).

Plate 6a: Showing eroded soil surface with maize crop

Plate 6b: Showing maize farm under severe sheet erosion

4.4 Tied ridging:

This practice is common in areas with insufficient or frequent prolonged dry spell occurrence during rain-fed cropping seasons. It involves construction of earthen bunds at right angle to the ridges at intervals of 1-2 m between any two ridges. Efficiency of tied ridge in conserving water depends on the following;

i). Rainfall distribution with respect to date of tied ridge.
ii). Distance of tied mound between the ridges
iii). Nature of soil
iv). Relief of the farm area

Rainfall distribution ought to be such that soil between the ridges does not become adversely ponded or waterlogged, but the water would infiltrate into the soil within a maximum of two days after last rain and before next rainfall event. Therefore, distance of mound tying would be such that would not break under the force or pressure of impounded water. Nature of the soil is an important consideration since clay soil would not be advisable for ridge tying. The practice would be suggested for loam soils or soils with good organic matter content to retain available water for crop use. The practice is also suggested for long season crops like cotton and sorghum (Planted in June and harvested in December to January) that require residual moisture for fruiting. Rainfall in the zone usually stops in September or very early October. Local relief of farm area is expected to be nearly flat to gently undulating to discourage erosion or runoff and allow water pressure on the mounds to be fairly even on the ridges. Ridging must be done against slope direction (Ogunwole and Odunze, 2002).

Tied-ridging used for soil moisture conservation

4.5 Planting pits:

Planting pit techniques are applicable in the arid zones or in areas having dominance of iron-hard set concretions that impair or does not make common tillage practices; like harrowing and ridging, profitable. Such areas would have very low water retention capacity, crusted soil surface that encourage runoff and discourage root growth and have shallow depth. The pits are often 20 to 30 cm diameter and 15 to 20 cm deep and are dug before rainfall establishment. Number of pit expected in one hectare of land could range between 12000 and 25000 depending on spacing between pits. However, the larger the pit diameter the more water can be retained/harvested in the pit (Ogunwole and Odunze, 2002). Soil is usually amended with farmyard manure and imported into the pit before planting is done. This practice is suggested for restoration of iron hard-set pans and crusted land for use in regular cultivation. The pits may however be surrounded with stones or earthen mounds to aid water harvesting.

4.6 Stone-bench or Mound Terracing:

Terracing is commonly practiced for hill slopes farming in the plateaus of Nigeria (Jos and Mambila) and is used to break hill slope length, control erosion and conserve soil moisture for crop use (Lontau *et al.*, 2002). In stone-bench terraced areas soil may be imported and kept in place with the stones arranged in bench forms (Plates 7 & 8) before planting of crops like sorghum (*Sorghum bicolor*), millet (*Pennisetum americanum*) or hungry rice (*Digitaria exils*). Imported soil is often amended with farmyard manure or household refuse to improve its fertility. Land preparation in such hill slope farming areas involve weeding, import of soil and organic manure and arranging stones on old terrace lines and making small depressions to store water for crop use (Plate 7b). Indigenous knowledge skills may account for the precision with which farmers place stones or earthen mounds in order to avoid erosion in their farms (Plates 7 & 8). In earthen mound terraced areas, crops like maize, Irish and sweet potatoes, and upland rice may be cultivated. Terraces in these areas are arranged into benches depending on the steepness of farmland slope (Plate 7a) and small depressions are made to keep water (Plate 7b) for roots' absorption.

Plate 7a: Showing stone terracing in the plateaus of Nigeria

Plate 7b: Showing stone terraced hill slope farming and provision for water storage for crop use in Jos Plateau of Nigeria

Plate 8a: Showing millet crop farming under hill slope terrace farming in plateaus of Nigeria

Plate 8b: Showing Millet crop grown in stone terraced hill slopes of Jos plateau of Nigeria

4.7 Cover cropping/Mulch Practices:

For the purpose of soil erosion control, conservation of soil moisture, improving microbiological activity in soil, improving soil quality and moderating soil temperature to enhance root zone biological activity for optimal crop production, agricultural soils are mulched either with straw (Dead) or planted (Live) mulch materials in the sub humid savanna of Nigeria (Plates 9 and 10). Straw from cereal crops could be cut into small pieces and spread uniformly on the field such that the soil surface is covered against direct impact of rain drop splash, water evaporation from the soil and control erosion (Plates 9a & b). This practice would not result in any significant competition between crop and mulch material. Rather, the mulch material would control soil erosion, modify soil temperature for enhanced soil biological activity, improve water infiltration, and on decomposition, improve soil quality, increase soil carbon and organic colloidal fraction for enhanced soil health and productivity. Plate 10 shows *Macrotyloma uniflorum* live mulch under-sown in a maize field. This technology require that the live mulch population and planting arrangement must not adversely affect main crop of the field, but would control erosion, modify soil temperature, improve infiltration rate of water, and impact on sustainable soil productivity. Extra advantage of this practice is that the mulch material could be used to raise seed or provide quality fodder for livestock, but such that livestock feed remains, urine and feaces are returned to the field during subsequent cropping season to further enhance soil quality/health for sustainable production.

Plate 9a: Showing straw (Dead) mulch practiced field crops

Plate 9b: Showing straw mulching under maize in an erosion control plot in Nigeria

Plate 10: Showing live mulching for erosion control in a maize field using *Macrotyloma uniflorum*

4.8 Residual Moisture Harvesting:

Residual soil moisture harvesting is commonly practiced in the sub humid savannas of Nigeria. The practice is used in the production of crops like tomatoes, cowpea, pepper, onion and garden egg (Longtau *et al.*, 2002). The practice is popular in upland areas where there are no nearby surface water bodies. It involves making ridges and planting or under sowing a cereal field in the month of August, weeding/pulverizing the soil before the last rainfall late in the month of September or early October and allowing the crop to extract residual moisture in the soil for establishment, flowering and fruiting. This practice has become very popular in providing tomatoes and pepper harvested and consumed in Nigeria; especially during the December festive periods (Christmas and Sallah), and many farmers have adopted the practice. The other implication of this practice is that it ensures that farmers intensively cultivate the land to have two harvests in one cropping season. The practice would therefore require carefully planned soil management practice in order to ensure sustainability of this productivity drive.

4.9 Irrigation/Supplemental irrigation:

Some crops are grown under supplemental irrigation in the sub humid savanna of Nigeria. Seasonal streams and inland valley water bodies provide water used for this practice. Crops grown under this practice include sugar cane, cowpea, tomatoes, pepper, onion, water melon, carrot, and egg plants, and are often planted in August or September. Supplemental water is supplied from late September up to harvest period in December/January. This practice is also very popular and farmers own water pumps (2 or 3 inch water pump engines) for lifting water and irrigating their fields by gravity method (Plates 11a, 11b and 11c).

Plate 11a: Showing water pump siphoning water from a seasonal stream for irrigating crops

Plate 11b: Showing irrigation pipe conducting water from stream to farm

Plate 11c: Showing water from stream through pipe irrigating grape vine stands in Nigeria

Plate 12a: Showing tomatoes field under irrigation using irrigation pump

Plate 12b: showing seasonal stream water used for irrigating Tomatoes and Cabbage fields

Plate 13: Showing cabbage field grown under irrigation using irrigation pump

The advantage of this practice is that farmers are able to obtain three crops within one season i.e., rain-fed crop, residual moisture crop, and supplemental irrigated crop and highly priced crops are focused. The practice appears to be a very intensive cropping management and should be backed with appropriate soil management and conservation practice(s). Some other crops are grown only under irrigation. These include grape vine orchards, rice (paddy and upland), wheat, vegetable maize, tomatoes, cabbage, spinach, onion, garlic, carrot, water melon and egg plants (Plates 12a, 12b, 12c & 13) . This practice has gained popularity among farmers that have facilities for extracting irrigation water from tube wells, seasonal inland valley water bodies/streams, and flood plains. Some adaptation of reservoirs using plastic water tanks to supply water for gravity irrigation is also in practice in some states of the sub humid Nigeria (Plates 14 &15)

Plate 14 Showing Plastic Water tank improvised as reservoir for irrigation water

Plate 15: Showing irrigation using bucket to supply water obtained from plastic tank

This adaptation has been used for growing highly priced crops like maize, and vegetables (carrot, tomatoes, water melon, cabbage and lettuce, onion and garlic). The devise (Plates 14 and 15) was used to grow maize (planted in February and harvested in May 2002 to 2005) at the 'Center For Energy Research and Training (CERT)', Ahmadu Bello University Zaria, Nigeria (Odunze *at al.*, 2008).

5.0 LIVESTOCK FARMING:

Livestock farming in Nigeria is still employing the free or unrestricted grazing approach. In this practice farmers take their herds to graze/browse fields without control (Plates 16a &b).

Plate 16a: Showing cattle grazing fields

Plate 16b: Showing cattle grazing cotton and maize fields

Usually, the herds graze harvested fields and from about December to May quality fodder becomes scarce. At this time farmers resort to feeding their herds with grass or let them out to browse any available plant material. The consequence of this include that crops yet to be harvested are grazed and fuels conflict between crop farmers and livestock herders, the soil is impoverished and degraded by animal trampling, animal health is impaired and farmer poverty situation is further aggravated. However, some conservation agriculture practice tried in the zone include

- ❖ Restricted grazing and Planned paddock ownership
- ❖ Crop-Livestock integration
- ❖ Organized grazing land

5.1 Restricted Grazing and Paddock Ownership:

The practice of restricted grazing with paddock ownership was advocated by the International Livestock Research Institute (Tarawali, 1992). The practice requires development of a paddock by livestock owner and planting the field with quality fodder materials like groundnut, *Centrosema pascuorum, Centrosema brasilianum. Macrotyloma uniflorum, Aeschynomenea histrix,* Stylosanthes species, *Lablab purpureus, Desmodium uncinatum* Ground nut, Soybeans and dual purpose cowpea or mixed grass and legume of choice. Fodder so generated will be harvested, cured, and fed to livestock as a supplemental feed during the feed scarcity periods of

December to May. This practice ensures that feed remains, urine and feaces of livestock is harvested, stored and returned to the field for incorporation at subsequent cropping seasons' land preparation. Also, livestock is fed quality fodder to ensure sustained animal health.

Plate 17: Showing herds of cattle out in field for uncontrolled grazing

5.2 Crop-Livestock integration:

Crop-livestock integrated farming practice provides that the crop farmer plans to produce grains for human consumption and commerce, as well as fodder for livestock and/or marketing. For example, Plate 18 shows groundnut haulms and grass bundles being marketed in Zaria for herders and livestock traders to buy. In this instance, farmers may grow dual purpose crops like maize, sorghum, groundnut and cowpea, harvest grains and crop residues. The residues are stored for later feeding of livestock. However, some farmers also grow fodder crops solely for feeding livestock and for sale at fodder scarcity periods (Plate 18). This practice implies that farmer can plan to produce required quality fodder for his herd and at least reduce adverse grazing of his livestock. In practice, this approach lends itself to planned investment on cultivable land for sustainable productivity improvement. Also, farmer could grow and market choice fodder from this system as a livelihood support entrepreneur. The system however requires that fodder is well cured to attract premium patronage and good price from customers.

Plate 18: Showing fodder display in a fodder market in Nigeria

5.3 Organized grazing land:

Planned grazing fields also exist in the sub humid savanna of Nigeria. In this practice grazing fields are developed by a community and are only grazed when permit is granted by designated authority. These grazing fields however appear insufficient to meet the feed requirements of the teaming livestock population that patronize such grazing fields. Therefore, fodder materials in such grazing fields could be bailed and sold to herders at affordable costs, or grazed on an agreed arrangement, but such that the fields are not adversely grazed by livestock.

5.4 Lessons Learnt:

Among lessons learnt in sub humid Nigeria conservation agriculture practice include the following:
- At land preparation, soil is not sufficiently moist to deter soil erosion by wind and contour bund and ridging has not been well adopted by farmers
- Hill slope farming is practiced in the zone but further support is needed to ensure sustainable crop production from such areas.
- Organic and inorganic fertilizer materials are not adequately used for crop production, perhaps further fertilizer use information dissemination would be necessary
- Soil conservation support practices like contour bund and ridging are not adequately practiced in the zone
- Human population increases in the zone has necessitated intensive cultivation of arable lands as available land is also put into residential, industrial, commercial and transport uses

➤ Livestock farming is not organized to ensure production of sufficient quality feed, protect and improve livestock health, and ensure sustainable soil productivity.
➤ Relevant Stakeholder institutions to support conservation agriculture in the zone exist

5.5 Stakeholders and Support:

Stakeholders in conservation agriculture in the Sub humid zones of Nigeria include agricultural research institutes i.e, Federal Universities of Agriculture in the Northern zones of Nigeria, like University of Agriculture Makurdi, Ahmadu Bello University Zaria, Federal Unversity of Technology Yola, University of Maiduguri, Maiduguri, Bayero University Kano, Usman Danfodio University Sokoto, Institute for Agricultural Research/Ahmadu Bello University Zaria National Animal Production Research Institute/Ahmadu Bello University, Zaria, National Agricultural Extension Research Liaison Services/Ahmadu Bello University Zaria, Colleges of Agriculture, International Institute of Tropical Agriculture, International livestock Research Institute, State Ministries of Agriculture and Agricultural Development Project in all Northern states in Nigeria, farmers, Agricultural support banks and the private sector investing on agricultural products (for exports, industrial, and merchandise). Supports from these organizations include:

- Technology acquisition and transfer
- Extension message delivery and farmer education
- Micro finance supports
- Information provision

6.0 MAIN CONSTRAINTS TO DEVELOPMENT OF CONSERVATION AGRICULTURE ACTIVITIES IN NIGERIA:

Main constraints to development of conservation agriculture in Nigeria would include the following:

I. Government Policy
II. Empowered Research Extension outfit
III. Farmer Empowerment

6.1 Government Policy:

The Federal Government of Nigeria is expected to provide enabling policy framework for conservation agriculture to be mandatorily adopted and practiced by farmers. This could be feasible through the Agricultural Research Council of the Nation providing acceptable guidelines for each articulated practice. For example, Policy on 'whole farm planning' could be developed for adoption by farmers. The main idea here is to enable farming communities develop conservation agricultural practices that maximizes economic returns from the land while ensuring that the soil is not degraded; in particular by erosion, nutrient mining by crops or agro-chemicals use in the process of farming. The system would involve an assessment of farm, taking into account issues like land area, soil types and topography, labour availability, machinery use, access to financial resources and farmer interest. The advantage of whole farm planning is that it provides farmers the opportunity to identify solutions for problems by themselves; in particular identifying conservation agriculture entry point in designated areas, identify individual farm or farm-level problem(s) and best solution. For example, early erosion occurrence in a particular farming community could best be identified by that farming community and with the assistance of Extension agents the problem could be solved before it escalates beyond the farming community's capacity to control. It may therefore be necessary to introduce 'whole farm planning' with some regulatory and financial support measures. In this instance, it would become a statutory requirement for farmers to implement farm plan or a specific action (E.g., CA entry point and Contour farming) once that action has been agreed by farmers in that community, in return for grant/ aid advanced to the community.

6.2 Empowered Research and Extension outfit:

The Nations' Research and Extension Institutions would need to be adequately empowered and financially mobilized to conduct research in conservation agriculture and to have the capacity to deliver and interact with farmers on research findings. For example, while the Institutes are funded, Research and Extension Scientists should develop research proposals that would be funded directly from the Agricultural Research Council of the Nation. These proposals should be in the areas of conservation agriculture that meet the Nations' projection of the **National Economic Empowerment and Development Strategies (NEEDS) programme and the Millenium Development Goals (MDGs)** i.e., to reduce poverty through enhanced agricultural productivity drive. Scientists involved in these studies should account directly to the National Agricultural Research council on funds granted and to Directors of their Institutes on research findings and suggest ways forward. Stakeholders from the private sector should be made to financially support conservation agricultural research and extension activities by policy since the life of their investments hinges largely on the success of conservation agriculture in the nation.

6.3 Farmer Empowerment:
Nigerian farmers are among the poorest of the world's populace, though they sustain the nation with provision of variety of food crops. Much more has to be done by the Nigerian Government and other stakeholders to alleviate the farmers' poverty situation. For example, farmers grow crops in impoverished soils, use little if any conservation agriculture practice, purchase fertilizer materials and agro-chemicals, pays for labour and contend with uncertain weather conditions unassisted. Yearly his yield returns depreciate substantially beyond a limit to sustain acceptable livelihood conditions. Sometimes he is compelled to abandon farming for any other vocation, even criminal, that presents ready option. Adequate conservation agriculture measures and insurance covers against farming risks should be provided for organized farmers. Regulations for conservation agriculture practice would need to be developed and rigidly followed by farmers. Credit, grant, and insurance organizations need to mobilize to access and disburse grants, aids and insurance coverage to farmer groups, while the issuing bodies must judiciously supervise to ensure fairness in funds administration, deployment by farmers and progress of venture being supported. Also, market information should be promptly disseminated to organized conservation agriculture farmers in each community to allow them take advantage of market situations to earn more from farm ventures to pay back to credit agencies terms of the credit and remain in a viable entrepreneur.

6.4 Suggested Areas of Research Focus
Suggested areas of research focus intervention for conservation agriculture in Nigeria include the following:
   1) Contour farming techniques for soil erosion control in conservation agriculture
   2) Water harvesting and conservation for sub humid areas
   3) Soil fertility Management for conservation agriculture
   4) Land use Management for conservation agriculture

6.5 Conclusion:

Everywhere in the world where people change a natural ecosystem into agriculture, the land degrades. The visible part is erosion, while other forms of degradation include chemical and biological forms. A number of factors are responsible for this degradation and include crop and livestock production on inappropriate land and use of inappropriate practice, overstocking and overgrazing of land, wrong timing of agricultural practices, uncontrolled tree felling, and lack of timely good vegetative ground cover provision. Conservation agriculture which aims at enabling farming communities develop agricultural practices that maximizes economic returns from the land while ensuring that the soil is not degraded in particular by erosion or agro-chemicals use in the process of farming is advocated. Conservation agriculture practice for each designated community should be well researched and packaged for farmer adoption under a researcher-farmer participatory arrangement. Government intervention with community level planned whole farm programme to enable farming communities develop conservation agriculture practices that maximizes economic returns from the land while ensuring that the soil is not degraded in particular by erosion or agro-chemicals use in the process of farming and provision of grants, insurance covers and aids for conservation agriculture farmers is strongly advocated.

6.6 References consulted:

Anthoni J. F. 2000. Soil erosion and conservation - contents index www.seafriends.org.nz/enviro/soil/erosion.htm

Dale Stephen, 1994. Preventing Gully Erosion in Nigeria. IDRC Project Number 921002] info@idrc.ca

Emechebe A.M.[1], B. James[2], T.K. Atala[3], I. Kureh[3], M.A. Hussaini[3], B.B. Singh[1], A. Menkir[4], A.C. Odunze[3], J.P. Voh[3], S.G. Ado[3] and S.O. Alabi. 2003. Farmer-Participatory On-farm Evaluation of *Striga hermonthica* Management Options in the Nigerian Northern Guinea Savanna Pilot Site of the SP-IPM. African Crop Science Conference, 2003, Nairobi, Kenya

Enwezor, W. O., E. J. Usoroh, K. A. Ayotade, J. A. Adepetu, V. O. Chude, and C. I. Udegbe. 1989. Fertilizer Use and Management Practices For Crops in Nigeria. Fertilizer Procurement and Distribution Division, Federal Ministry of Agriculture, Water Resources and Rural Development. Lagos. P 163

Inman Alex, 2006. Soil Erosion in England and Wales: cause, consequences and policy options for dealing with the the problem. Discussion paper prepared for WWF. WWF-UK,Panda House, Weyside Park, Godalming, Surrey GU7 1XR www.wwf.org.uk

Jones, M. J. and A. Wild. 1975. Soils of West African Savanna. The maintenance and improvement of their fertility. Commonwealth Bureau of Soils. Technical Communications № 55 CAB Harpenden.

Lal,R. 1975. Role of mulching techniques in tropical soil and water management. Technical Bulletin No.1 International Institute of Tropical Agriculture (IITA), Ibadan, Nigeria

Longtau S, Odunze A.C, and Ben Ahmed. 2002. Case Study of Soil and Water Conservation (SWC) in Nigeria. Field Survey Reports. In Rethinking Natural Resource Degradation in Sub-Saharan Africa: Policies to support sustainable Soil Fertility Management, Soil and Water Conservation among resource-poor farmers in Semi-arid Areas. (Tom Slaymaker and Roger Blench eds.) Overseas Development Institute (UK) & University for Development Studies. Tamale, Ghana. Vol. 11: Chapter 111:1-41.

Lombin, G. 1987. Soil and Climate Constraints to Crop Production in the Nigerian Savanna Region. 15th annual conference of Soil Science Society of Nigeria Proceedings. Kaduna, Nigeria. 21st- 24th September 1987

Odunze, A. C. 2003. Northern Guinea Savanna of Nigeria and Rainfall Properties for Erosion Control. African Soils / Sols Africans Journal. Vol. 33:73-81

Odunze, A. C., S. A. Tarawali, N.C. de Haan, E. N. O. Iwuafor, P. D. Katung, G. E. Akoueguon, A. F. Amadji, R. Schultze-Kraft, T. K. Atala, B. Ahmed, A. Adamu, A. O. Babalola, J. O Ogunwole, A. Alimi, S. U. Ewansiha, and S. A. Adediran. (2004). Grain Legumes for Soil Productivity Improvement in the Northern Guinea Savanna of Nigeria. Food, Agriculture, and Environment Journal. Vol. 2(2): 218-226. Helsinki Finland

Odunze A.C and J. O. Ogunwole. 2002. Proven On-Farm Soil Conservation Technologies in the sustenance of Soil Fertility in Northern Nigeria. . In: Proven Technologies for Soil Fertility Management. Proceedings of a Workshop on Inventorization of Soil Management Technologies for Increased Productivity. Chude, V. O., C. O. Ezendu, S. A. Ingawa, and O. O. Oyebanji (eds.). National Special Programme For Food Security (NSPFS), Abuja, Nigeria, and Food and Agriculture Organization of the United Nations (FAO). Pp 41-57.

Odunze A. C. Jinshui Wu, Liu Shoulong, Zhu Hanhua, Ge Tida, Wang Yi and Luo Qiao. (2012). Soil Quality Changes and Quality Status: A case study of the Subtropical China Region Ultisols. British Journal of Environment and Climate Change. SCIENCEDOMAIN INTERNATIONAL, Vol. 2(1) 37-57, 2012

Odunze, Azubuike Chidowe, Ebireri, Onome Felicia, Ogunwole, Olalekan Joshua, Tarfa, Bitrus Dawi and Eche, Nkechi Mary. (2013). Effect of Tillage and Fertilizer on Soil Quality and Yield of Maize in an Alfisol of a Northern Guinea Savanna of Nigeria. Journal of Agriculture and Biodiversity Research, Vol. 2(8): 167-177. October 2013.Available online at http://www.onlineresearchjournals.org/JBAR

Odunze Azubuike Chidowe (2015). Soil Conservation for Mitigation and Adaptation to a Changing Climate: Sustainable Solutions in the Nigerian Savanna Ecology. International Journal of Plant & Soil Science; 8(4): 1-12, 2015; Article No. IJPSS 19628, SCIENCEDOMAIN International. www.sciencedomain.org

Odunze, A.C., K.B. Adeoye, J.J. Owonubi, V.O. Chude, E.N.O. Iwuafor and J.K. Adewumi (1997). Rainfall characteristics and soil tillage timing for rainfall crop production in the Northern Guinea Savanna of Nigeria. The ITC Journal Netherlands. Vol. 3/4 (In CD-ROM); 59-65

Odunze A. C and I Kureh (2007). Land use Limitations and Management Option for a Savanna zone Alfisol. Nigerian Journal of Soil and Environmental Research Vol.7: 70-81 (2007)

Odunze, A. C., L. A. Dim and L. K. Heng (2008). Water Supply and Rain-fed Maize Production in a Semi-Arid Zone Alfisol of Nigeria. 2008 International Soil Conservation Organization Conference Paper. Buderpest Hundary

Ogunwole, J. O., and Odunze A. C. 2002.Proven On-Farm water harvesting for rainfed crops in Nigeria. . In: Proven Technologies for Soil Fertility Management. Proceedings of a Workshop on Inventorization of Soil Management Technologies for Increased Productivity. Chude V. O., C O. Ezendu, S. A. Ingawa, and O. O. Oyebanji (eds.). National Special Programme For Food Security (NSPFS), Abuja, Nigeria, and Food and Agriculture Organization of the United Nations (FAO). Pp 101-110.

Odunze A. C. (2011). Use of Meteorological Data in Agriculture. In Training Manual on "Training of Monitoring and Evaluation Staff Handling Weather Data Collection of Kaduna State Agricultural Development". Farming Systems Research Programme, Institute for Agricultural Research/Ahmadu Bello University, Zaria. FRSP Special Publication. Pp 1-9

Tarawali, G. (1991). The residual effect of Stylosanthes fodder banks on maize yield at several locations in Nigeria. Tropical Grasslands 25: 26-31

Tarawali .S. A., **Odunze, A. C.,** Firmin, A., de Haan, N. C., Iwuafor, E. N. O., Atala, T. K., Ahmed, B., Adamu, A., Katung, P. D., Akouenon, G. E., Owoeye, L. G., Ajaero, J., and Ewansiha, S.U. 2003. Development of Strategies to Promote Farmer Utilization of Herbaceous Legumes for Natural Resource Management to Improve Farm Income and Food Security. A Collaborative project between BMZ/GTZ of Germany, IITA and ILRI, Ibadan, Nigeria and IAR/ABU. Zaria, Nigeria. Final Report. International Institute of

Tropical Agriculture (IITA) Ibadan, Nigeria

7.0 SUGGESTED CONSERVATION PRACTICES FOR RICE AND CASSAVA PRODUCTION IN NIGERIA

Suggested practices for rice and cassava production packages employing Conservation agriculture and suitable for the Nigerian situation are discussed in the next sections.

7.1 Soil Fertility and Quality Management for Rice (Oryza sativa L.) Production
7.1.1 Site selection and land preparation.

Basically, two types of rice varieties based on ecology are grown; viz. Upland and Lowland rice. Upland rice is grown under rain-fed, free draining soil conditions. Preferably, site for this should be on a flat to very gently sloping terrain. Land preparation will involve ploughing, harrowing and basin construction. If site is on a gently sloping terrain (slope $\leq$ 8%), contour banding will be necessary before basins are constructed along the contours. Field drainage channels must be built and maintained to ensure that excess salts are conducted away from the field against salt build-up. Also sufficient water ($\geq$5cm water head) must be supplied in the basins to allow for leaching of excess salts and conducted away in the drainage channels. This is to ensure that soil health is maintained and erosion is not encouraged. Crop residues must be incorporated at land preparation and herbaceous legumes; such as *Centrosema brasilianum, Pueraria phaseolus and Mucuna* could be relayed on the field soon after draining water off the basins for rice harvest. Land rotation should be built into the Land preparation plan; such that at least, a one year planted fallow with herbaceous legumes is accommodated and the legume incorporated at land preparation in subsequent cropping season. Herbaceous legumes for the one year planted fallow could include *Centrosema pascuorum, Macrotyloma uniflorum,* and *Stylosanthes hamata*. Burning of rice wastes and residues is discouraged. Waste rice biomass after harvest should be chopped into smaller bits and incorporated into the soil at land preparation to improve and maintain soil quality/health and fertility for sustainable soil productivity.

Lowland rice varieties are grown on flood plains, irrigated schemes and deep flooded areas. At the seasonally wet lands, land clearing, ploughing and harrowing should be done before the soil is wet with mechanized equipment. Scrapping of top soil, removal of plant roots, stumps must be avoided at land clearing and drainage channel must be built into the field at land preparation. Basin construction and puddling with mechanized tools should be done after harrowing and when sufficient water (≥5cm water head) is available in the basins. A planted short fallow with herbaceous legumes like *Desmodium uncinatum, Pueraria phaseoloides,* and *Centrosema brasilianum* to be relayed at drainage before harvest and incorporated at subsequent land preparation should be accommodated.

7.1.2 Varieties:

It is important that appropriate rice variety for the ecology is planted if planned yield level is expected. Upland rice varieties recommended for the various agro-ecological zones of Nigeria are shown in Table 1. The National Institute of Cereals Research (NCRI) Badegi and the Africa Rice Project are actively breeding improved rice varieties for both Upland and Lowland rice production in Nigeria. The Agricultural Extension Agents (EAs) should advice farmers on current best varieties for the ecology.

Table 1: Recommended upland rice varieties for the different agro-ecological zones

AGRO-ECOLOGICAL ZONE	RECOMMENDED UPLAND RICE VARIETY
Sahel	FARO 45, FARO 46 EX-China, FARO 55 (NERICA
Sudan	FARO 45, FARO 46, EX-China, FARO 38, FAR FARO 55 (NERICA 1)
Northern Guinea Savanna	FARO 46, FARO 39, FARO 38, FARO 11, FAR FARO 55 (NERICA 1), FARO 56 (NERICA 2) FA (NERICA 7), FARO 59 (NERICA 8), FARO 62 (O 1), FARO 63 (OFADA 2)
Southern Guinea Savanna	FARO 46, FARO 48, FARO 49, FARO 43, FAR FARO 55 (NERICA 1), FARO 56 (NERICA 2) FA (NERICA 7), FARO 59 (NERICA 8), FARO 62 (O 1), FARO 63 (OFADA 2)
	FARO 46, FARO 48, FARO 49, FARO 43, FAR

Forest	FARO 55 (NERICA 1), FARO 56 (NERICA 2) FA[RO] (NERICA 7), FARO 59 (NERICA 8), FARO 62 (O[FADA] 1), FARO 63 (OFADA 2)

For Lowland rice varieties, Table 2 presents a list of recommended rice varieties for each agro-ecological zone in Nigeria.

Table 2: Recommended lowland rice varieties for different agro-ecological zones

AGRO-ECOLOGICAL ZONE	RECOMMENDED LOWLAND RICE VARIETY
Hydromorphic and inland valley swamp	FARO 44, FARO 52, FARO 31, FARO 15, FAR[O] FARO 51 FARO 62 (OFADA 1), FARO 63 (OFA[DA] FARO 60 (NERICA L19), FARO 61 (NERICA L34)
Shallow swamp and irrigated swamp	FARO 44, FARO 52, FARO 51, FARO 27, FAR[O] FARO 37, FARO 60 (NERICA L19), FARO 61 (N[ERICA] L34)
Deep water and floating	FARO 15, CK 73, DA 29, BKN 6986 – 17, ROK 5, I[R]
Mangrove	FARO 15, ROK 5, WAR 77-3-2-2, FARO 28, IR 54

7.1.3 Weed control:

The following are herbicides recommended for weed control in upland rice production and their rates of application.

(i)   Propamil + oxadiazon at 3.0kg a.i. ha^{-1} (5 liters Ronstar 400 EC/ha) or
(ii)   Glyphosate e.g Roundup (4 - 6 litres ha), 2 weeks before planting followed by either
(iii)   Propamil + bentazon at 3.0kg a.i. ha^{-1} or
(iv)   Propamil + Fluorodifen at 3.0kg a.i. ha^{-1} or
(v)   Propamil + thiobencarb at 3.0kg a.i. ha^{-1} 2 - 3 weeks after planting
(vi)   Butachlor at 4l/ha pre-emergence
(vii) Oxadiazon at 4 – 5l/ha pre-emergence but must be applied at least a week before transplanting of rice

(viii)   Propanil at 4l/ha Post-emergence

(ix)   Propanil + Triclopyr at 4l/ha Post-emergence.

For lowland and upland rice weed control, the under listed herbicides are recommended for effective weeds control in rice production.

x. Propamil + oxadiazon at 3.0kg a.i. ha^{-1} (5 liters Ronstar 400 EC/ha) or
xi. Glyphosate e.g Roundup (4 - 6 litres ha), 2 weeks before planting followed by either
xii. Propamil + bentazon at 3.0kg a.i. ha^{-1} or
xiii. Propamil + Fluorodifen at 3.0kg a.i. ha^{-1} or
xiv. Propamil + thiobencarb at 3.0kg a.i. ha^{-1} 2 - 3 weeks after planting
xv. Butachlor at 4l/ha pre-emergence
xvi. Oxadiazon at 4 – 5l/ha pre-emergence but must be applied at least a week before transplanting of rice
xvii. Propanil at 4l/ha Post-emergence
xviii. Propanil + Triclopyr at 4l/ha Post-emergence

7.1.4 Cultural Practices:

For lowland rice use 60 - 70kg ha^{-1} seed, drilled directly on flat 20cm x 20cm, or seed 40 - 45kg ha^{-1} for nursery bed and then transplant 1 - 2 seedlings per hill spaced at 20 x 20cm at 21 – 28 days old seedling or use 50 - 60kg ha^{-1} seed for direct seeding. For deep water rice use 40 - 45kg/ha seed broadcast or drill on the flat at 30 x 30cm. For both upland rice and direct seeded lowland rice, plant when the rains are regular or sufficient water is available. Note that seeds for planting should be treated with seed treatment chemicals such as Apron Stat or seed plus to control insects and fungal attack.

7.1.5 Fertilizer use, type and rate:

i). For lowland rice (shallow swamp, irrigated, hydromorphic and inland valley swamp) apply half the N and all P and K at planting/transplanting and the remainder broadcast at 6 - 7 weeks after planting/transplanting or at panicle initiation stage.

ii). for lowland rice (deep water and floating and mangrove ecologies), apply all N, P and K at planting.

(iii). For upland rice in Sahel, Sudan and Northern Guinea, apply half N and all P and K at 1 - 2 weeks after planting, band apply or broadcast the remainder of N at 6 weeks after planting and when sufficient water (≥5cm head) is available in the basins.

iv). For upland rice in Southern Guinea and Forest zones, apply all N, P and K. 1 - 2 weeks after planting and first weeding

Fertilizer sources and rates recommendations for upland and lowland rice (based on soil test/soil fertility map) are presented below.

Fertilizer sources and rates

100kgN	20:10:10	5 bags basal
	Urea	2 bags topdress
80kgN	20:10:10	4 bags basal
	Urea	1½ bags topdress
60kg N	20:10:10	3 bags basal
	Urea	1½ bags topdress
40kg N	20:10:10	2 bags basal
	Urea	1 bag topdress
100kgN	15:15:15	7 bags basal
	Urea	2 bags topdress
80kgN	15:15:15	5 bags basal
	Urea	1½ bags topdress
60kg N	15:15:15	4 bags basal

	Urea	1¼ bags topdress
40kg N	15:15:15	3 bags basal
	Urea	1 bag topdress

Depending on the fertility class of farmland, Table 3, could guide the fertilizer rates of application for sustainable upland and lowland rice cultivation.

Table 3: Fertilizer recommendations for upland and lowland rice (based on soil test/soil fertility map)

NUTRIENT	FERTILITY CLASS	UPLAND RICE	LOWLAND RICE
N	Low	80kg N	100kg N
	Medium	60kg N	80kg N
	High	40kg N	40kg N
P	Low	30 - 40kg P_2O_5	40 - 50kg P_2O_5 "b"
	Medium	30kg P_2O_5	40kg P_2O_5
	High	NIL	NIL
K	Low	30 - 40kg K_2O	30 - 40kg K_2O
	Medium	30kg K_2O	30kg K_2O
	High	NIL	NIL

7.1.6 Nutrient Deficiency/Toxicity Symptoms:

It is a common experience to notice deficiency symptoms in rice fields. Common among them are presented below

i). *Nitrogen Deficiency:* Characterized by stunting, and poor tillering. Leaves are narrow, short, erect and yellowish-green. Old leaves die when straw is coloured.

ii). *Phosphorus Deficiency:* Plants are stunted with a limited number of tillers. Leaves are narrow, short, erect and dirty-dark green. Old leaves die when brown coloured. A reddish or purplish colour may develop on leaves if the particular rice variety has a tendency to produce anthocyanin pigment.

iii). *Potassium Deficiency:* Stunted and weak plants. Leaves short, droopy and dark green. Sometimes; brown spots may develop on the dark green leaves.

iv). *Magnesium deficiency:* With mild deficiency, no clear-cut symptoms more severe deficiencies usually cause wavy and droopy leaves. Inter-venial chlorosis occur on lower, leaves, sometimes, characterized by orange yellow colour.

iv). Sulphur Deficiency: Yellowish colouration of young leaves.

v). *Zinc Deficiency:* The more common symptoms are the appearance of brown blotches, streaks on the lower leaves, followed by stunted growth. In the field, uneven growth and delayed maturity are characteristics of Zn deficiency.

vi). *Iron Deficiency:* Uppermost leaves of the plant become chlorotic with some green colour retained around the veins. The young leaves take on a bleached appearance. Older leaves retain their green colouration at first, but as the deficiency progresses, they become chlorotic with marked interveinal chlorosis. Iron deficiency in rice occurs in irregular patches in the field; green and chlorotic stands have often been seen to grow side by side. Iron deficiency is often associated with soils high in pH (e.g saline -sodic, sodic soils and Vertisols)

vii). Iron toxicity: At first, yellowing and tiny brown spots appear on the lower leaves starting from the tips and spreading towards the base. Subsequently, younger leaves become affected and many older leaves completely die. In susceptible cultivars, the leaf colour may be orange, yellowish - brown, reddish - brown, brown or purplish brown, depending on the variety and severity of iron toxicity. Roots of affected plants are generally coarse, sparse dark brown and damaged

Farmers and EAs should work in close collaboration to mitigate nutrient deficiency and toxicity on rice crops in the field in order to avoid yield depletions at harvest, by promptly reporting any identified symptom(s) to supervising research outfits in the agro-ecological zone.

7.1.7 Pests and Disease incidence:

Incidents of pests and disease occurrence, as well as birds' incursions on rice field should be promptly guarded against or reported for prompt action. However, the practice of placing field nests over the rice plants to shield the grains away from birds is encouraged.

7.1.8 Yield Expectancy:

Farmers' yields currently range between 1.2 and 3.0tha^{-1} for swamp rice and 1.0 – 1.5tha^{-1} for upland rice. With improved management practices yields of up to 5.0 – 6.0tha^{-1} and 2.5 – 3.0tha^{-1} of paddy are possible for swamp and upland rice, respectively.

7.1.9 Agricultural Extension Agents (EAs):

To ensure that rice production is cost effective and a dependable entrepreneurial venture, the services of EAs to farmers must the strengthened. To achieve this, the State Agricultural and Rural Development Authorities must work in concert with the farming communities to provide sufficient number of EAs, mobilize them appropriately and monitor performance of the EAs, the farmer and the farm venture. This will give confidence to the EAs; especially when well remunerated, and the farmer to continue in the production venture.

Agricultural Extension Service should also source for markets for the farm produce and products, link farmers to markets, credit agencies, viable processing outfits and form farmers into cooperatives to aid access of credit facilities.

Government efforts to buy off excess rice, bulk process, package and conserve in appropriate storage facility (ies) as a national food security strategy is encouraged in order not to discourage the farming community who may not exhaust locally the rice produced in any one year, as well as maintain on a sustainable basis, rice production entrepreneurship to contribute to the national Gross domestic Product (GDP).

7.1.10 References consulted

Chude, V. O., S. O. Olayiwola, A. O. Osho and C. K. Daudu (2011). Fertilizer Use and

Management Practices for Crops In Nigeria. 4th Edition. Federal Fertilizer Department Federal Ministry of Agriculture and Rural Development, Abuja

Ethan, S., **A. C. Odunze**, S. T. Abu and E. N. O. Iwuafor (2011). Effect of Water Management and Nitrogen Rates on Iron Concentration and Yield in Lowland Rice. Agriculture and Biology Journal of North America 2011, 2(4): 622-629

Ethan S., **Odunze A. C**, Abu. S. T, and Iwuafor E. N. O (2012). Field Water Management and N-Rates to Save Water and Control Iron Toxicity in Lowland Rice. Academic Research International Journal Vol. 3. No. 2 September 2012: 184-201

Ezui, K.S., C.K. Daudu, A. Mando, M.T. Kudi, **A.C. Odunze**, J.O. Adeosun, I.Y. Amapu, B. Tarfa, I. Sambo, I. Bello and C. Dangbegnon (2010). Informed site-specific fertilizer Recommendation for upland rice production in northern guinea savannah of Nigeria. Second Africa Rice Congress 2010 Proceedings. Innovation and Partnerships to realize Africa's Rice Potential 22-26th March 2010 Bamako, Mali

Kudi, T. M, Daudu, C. K, Ezui, K. S, Mando, A, Adeosun, J. O., **Odunze, A. C**, Amapu, I. Y., Sambo, I. J., Dangbegnon, C., Bello, I. and Ayoola, J.B. (2011). Profitability of best bet Fertilizers options towards improved rice productivity in Northern Guinea Savannah Zone of Nigeria. Journal of Research in Agriculture, Vol. 8(3): 37-41

Odunze A. C., M. T. Kudi, C.K. Daudu, J. Adeosun, G. Ayoola, I. Y. Amapu, S. T. Abu, A. Mando, G. Ezui, and D. Constance. 2010. Soil Moisture Stress Mitigation for Sustainable Upland Rice Production in the Northern Guinea Savanna of Nigeria. Journal of Development and Agricultural Economics Vol. 2(11) 382-388

7.2 Soil Quality and Fertility Management for Cassava Production under Conservation Agriculture:

7.2.1 Introduction:

Cassava *(Manihot esculenta Crantz)* is an important food crop both for urban and rural consumers in Africa

(FAO online). In recent years, cassava is increasingly gaining importance as a cash crop for small-scale farmers in

Nigeria in particular. Nigeria is currently the major cassava producing nation; accounting for 35% worldwide cassava production (FAO, 2005). However average yield of cassava has remained below 10tha^{-1}, while annual growth rate of 2% is unabated. Also in Nigeria, land under agricultural production use is currently under intensive exploitation with resulting decrease in soil quality, fertility and consequent productivity. Efforts at conserving, restoring, maintaining soil conditions for sustainable productivity while exploiting this nonrenewable natural resource must be intensified in order to attain yield levels beyond 15tha^{-1} cassava production on a sustainable basis.

Site selection:

Choose an accessible well-drained fertile soil. Bear in mind that cassava is a tuber and under poorly drained soil, the soil could be acid to cause tuber decay; resulting in reduced tuber yield. Also consider the slope of your selected field to be sure that it is not prone to erosion. If field is on a gently sloping terrain (slope $\leq 8\%$), contour ridging would be advised. Bear in mind that erosion will wash out the soil and nutrients and degrade the land for use in sustainable agricultural production, especially on a strongly sloping field. Consult your Agricultural Extension Agent (EA) for advice on appropriate conservation practice for sustainable cassava production at this site. Conservation Agriculture practices are often site-specific for purposes of appropriate use as intervention measure(s).

7.2.3 Varieties

The following varieties are recommended for their high yield and processing quality: TMS 30572, NR 8082, NR8083, TMS 4(2) 1425, TMS 81/00110, TMS 92/0326. However, consult your Agricultural Extension Agent for information on the most suited variety for your ecology (in terms of yield and processing quality). The Nigerian Root Crops Research Institute (NRCRI) Umudike and the International Institute of Tropical Agriculture (IITA) Ibadan are actively engaged on research to provide improved cassava varieties with high yielding and processing quality potentials. Exploit their services to improve your cassava venture through the EA in your location. Table 1 shows a list of other cassava varieties recommended for States across the country from which you may choose.

Table 1: Recommended cassava varieties for different zones of Nigeria

South-Eastern states includes Abia, Akwa-Ibom, Anambra, Bayelsa, Cross River, Ebonyi, Imo, Rivers	NR 8082, NR 8083 TMS 30572, TMS 30555,MS 4(2) 14 Nwugo
South-Western states including Delta, Edo, Ekiti, Kwara, Lagos, Ogun, Ondo, Olsun, Oyo	TIMS 30572, NR 8082, NR 8083
Northern States including Adamawa, Bauchi, Benue, Borno, Gombe, Jigawa, Kaduna, Katsina, Kebbi, Kogi, Nasarawa, Niger, Plateau, Sokoto, Taraba, Zamfara, FCT.	TMS 30572 4 (2) 1425 NR 8082, NR 8083

7.2.4 Weed control and land preparation:

A total herbicide – Round up (glyphosate) should be applied at the rate of 4–5 l/ha 10 days before land preparation. Where a total herbicide was not used before land preparation, it would be recommended that a selective pre-emergence herbicide be applied within three days after planting. Five litres of Primextra is recommended /ha. Table 2 presents other herbicides recommended for weed control in cassava production.

Table 2: Herbicides recommendation for cassava

HERBICIDE APPLICATION	RATE OF APPLICATION a.i. (kg ha)	TIME OF
Primetra	3.2	Pre-emergence
Gramoron + Paraquant	1.4, 2.5	Pre- emergence
Fluometeron + paraquant	3.0, 2.5	Pre- emergence

For cost effectiveness and optimum plant population, mechanization and planting on ridges are recommended. Do not ridge your field along the up-down slope direction so as not to conduct soil and water at erosive speed away from your farm. This is one common farmer practice that has degraded the soil quality and fertility over time.

During land preparation, ensure that previous year relayed herbaceous legumes in-situ grown and residues are incorporated into the soil. Also prefer to use mechanized implements to plough, harrow and ridge on flat farmlands; especially for farmland $\leq$1.0ha, while employing reduced tillage Conservation Agriculture practices. Legumes that could be tried for in-situ grown intercrop arrangements include *Centrosema pascuorum, Macrotyloma uniflorum* and *Desmodium uncinatum*. For relay planting and/or short planted fallow practice, *Centrosema brasilianum, Mucuna* and *Pueraria phaseoloides* could be tested f or environmental suitability and socio-economic preference of the people. Crop rotation practices involving legume sole and legume intercrops; such that will not depress yield of cassava tuber, is recommended. Soil is a non renewable natural resource. All efforts at conserving the soil must be undertaken while exploiting it for sustainable agri production enterprise.

7.2.5 Planting and planting material:

1. Planting starts in February/March and can be extended to October depending on the agro-ecology of your location. However, early planting is advised. Liaise with your EA for appropriate cassava variety/planting material for an entrepreneurship relative to your ecology and time of planting in the year.
2. The quantity recommended for 1 ha is 60 bundles of cassava
3. Stem cuttings 25 cm long should be planted at a spacing of 1m x1m

4. Maintain 100% planting rate by replacing dead or nonviable stems.

7.2.6 Fertilizer use, type and rate

For sustainable cassava production, the following fertilizers and their rate/ha are recommended

- NPK 15:15:15–12 (50 kg)
- NPK 20:10:10–9 (50 kg)
- NPK 12:12:17–15 (50 kg) bags

Table 3 presents a list of other recommended fertilizer types and rates/ha for cassava production.

Table 3: Other recommended fertilizers for cassava

NUTRIENTS/ha	MATERIALS ha^1	RECOMMENDATION
15-15-15	12 bags	600kg ha^{-1}
20-10-10	9 bags	450kg ha^{-1}
12-12-17+2Mg0	15 bags	750kg ha^{-1}

Apply fertilizer at 8 weeks after planting. Apply fertilizer in a ring, 6 cm wide and 10 cm from the plant and bury the fertilizer. Do not broadcast fertilizer even around the plant so that you do not lose the nitrogen largely to volatilization or surface wash erosion. Ensure that the fertilizer does not touch the stem or leaves. Under intensive cultivation practice, relay appropriate herbaceous legume; such as *Centrosema pascuorum*, *Macrotyloma uniflorum*, *Desmodium uncinatum* etc) at 6 weeks after planting cassava on one side of the ridge slope to conserve soil moisture, check soil loss to erosion, improve soil quality and fertility for sustainable productivity of the soil and provide feed for livestock. Also plan a crop rotation arrangement at which a short fallow with herbaceous legume; like *Centrosema brasilianu*, *pueraria phaseoloides* and Mucuna could either be relayed at cassava harvest to last one year and be incorporated subsequently or planted sole over a one year period to improve the soil quality and fertility for sustainable cassava production.

As a complimentary soil fertility management measure, 3 t of poultry manure + 200 kg NPK (20. 10.10) per hectare annually would maintain and restore soil quality and fertility for sustainable cassava production

7.2.7 Yield Expectancy

Yield of over 15t/ha can be obtained with good agronomic practices and soil management. As a working tool to guide your cassava production venture, Table 4 attempted a rough estimation of cost of cassava production on a one hectare (1ha) land area.

Table 4: Production cost for one hectare of cassava to ensure yield of over 15 t/ha

		N
1	Land preparation	10,000.00
2	Cassava cuttings (60 bundles @ N300/bundle)	18,000.00
3	Planting (8 pd* at N500/pd)	4,000.00
4	Pre-emergence herbicides (5 liters at N1000/l)	5,000.00
5	Fertilizer (20:10:10, 9 bags at N2500/bag)	22,500.00
6	Insecticides (2 liters at N1000/l)	3,000.00
7	Application of herbicide	3,000.00
8	Application of insecticide	3,000.00
9	Application of fertilizer (8 pd at N500/pd)	4,000.00
10	One weeding (20 pd at N500/pd)	10,000.00
11	Harvesting (35 pd at N500/pd)	17,500.00
	Total	**100, 000.00**

*pd= person days. Farm labor wage rates vary by location

Note: Fixed capital investments are not included. Such capital investments include knapsack or boom sprayers, tractors or power tillers, stem cutters, planters, and harvesters. With planters and harvesters, manual labor use can be minimized.

7.2.8 Agricultural Extension Agents (EAs):

To ensure that cassava production is cost effective and a dependable entrepreneurial venture, the services of EAs to farmers must the strengthened. To achieve this, the State Agricultural and Rural Development Authorities must work in concert with the farming communities to provide sufficient number of EAs, mobilize them appropriately and monitor performance of the EAs, the farmer and the farm venture. This will give confidence to the EAs; especially when well remunerated, and the farmer to continue in the production venture. Agricultural Extension Service should also source for markets for the farm produce and products, link farmers to markets, credit agencies, viable processing outfits and form farmers into cooperatives to aid access of credit facilities. Government efforts to buy off excess cassava tuber, bulk process into various products and conserve in appropriate storage facility (ies) as a national food security strategy is encouraged in order not to discourage the farming community who may not exhaust locally the cassava produced in any one year, as well as maintain on a sustainable basis, cassava production entrepreneurship to contribute to the national gross domestic product (GDP).

7.2.9 References Consulted:

Chude, V. O., S. O. Olayiwola, A. O. Osho and C. K. Daudu (2011). Fertilizer Use and Management Practices for Crops In Nigeria. 4th Edition. Federal Fertilizer Department Federal Ministry of Agriculture and Rural Development, Abuja

Daniel K. Adedzwa, **A.C. Odunze,** A.F. Lum; Ayitope Aremu; Ntekim J.; Ben Odoemena; J. Eroh; N Nwachukwu; and J. Ajaero (2004). Community Analysis of Ikot Etuk Udo (Abak Urban), Abak Local Government Council, Akwa Ibom State, Nigeria. Final Report. IITA document.

FAOSAT (Online statistics http://faostat.fao.org/site/567/desktopDefault.aspx?PageID=567#anchor)

FAO. Global Forest Resources Assessment (2005). Food and Agriculture Organization of the United Nations: Rome, Italy 2005.

Odunze, A. C, Ajaero, J. O., Odeh J. O, Eroh, J. O. and Nwachukwu, N.2005. Cassava Enterprise Development (CEDP) Community Analysis for Cross River State Nigeria (Ukim Ita Village). Final Report. IITA Document Cassava production Internet document.

Amadi, C. O, Ekwe, K. C, Chukwu, G. o, Olojede, A. O and Egesi, C. N (2011). Root and Tuber Crops Research for Food Security and Empowerment. Published by National Root Crop Research Institute (NRCRI) Umudike, Nigeria. P714, www.nrcri.gov.ng

I want morebooks!

Buy your books fast and straightforward online - at one of the world's fastest growing online book stores! Environmentally sound due to Print-on-Demand technologies.

Buy your books online at
www.get-morebooks.com

Kaufen Sie Ihre Bücher schnell und unkompliziert online – auf einer der am schnellsten wachsenden Buchhandelsplattformen weltweit!
Dank Print-On-Demand umwelt- und ressourcenschonend produziert.

Bücher schneller online kaufen
www.morebooks.de

OmniScriptum Marketing DEU GmbH
Heinrich-Böcking-Str. 6-8
D - 66121 Saarbrücken
Telefax: +49 681 93 81 567-9

info@omniscriptum.com
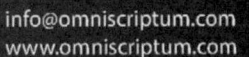

Printed by Books on Demand GmbH, Norderstedt / Germany